SPOTLIGHT ON NATURE

CROCODILE

LAURA PURDIE SALAS

CREATIVE EDUCATION · CREATIVE PAPERBACKS

Published by Creative Education and Creative Paperbacks
P.O. Box 227, Mankato, Minnesota 56002
Creative Education and Creative Paperbacks are imprints of The Creative Company
www.thecreativecompany.us

Design by Jen Bowers; production by Beeline Media & Design, Inc.
Art direction by Tom Morgan
Printed in the United States of America

Photographs by Getty Images (Cathy Scola), Shutterstock (Stephen Barnes, Rattanasak Khuentana, Mark_Kostich, reptiles4all, Nikolay 007, Alessandro De Maddalena, jeep2499, PUMPZA, Kurit afshen, Lauren Suryanata, Vladimir Turkenich, Zeno Swijtink, Cheryl Jayaratne, Boris Mayer, Sergey Uryadnikov, jo Crebbin, Eric Isselee, Volodymyr Burdiak, wilaiwan jantra, fotoslaz, JTKP, Matyas Rehak, Tanyapatch, You Touch Pix of EuToch, Padodo)

Library of Congress Cataloging-in-Publication Data
Names: Salas, Laura Purdie, author.
Title: Crocodile / Laura Purdie Salas.
Description: Mankato, Minnesota : Creative Education and Creative Paperbacks, [2025] | Series: Spotlight on nature | Includes bibliographical references and index. | Audience: Ages 10-13 | Audience: Grades 4-6 | Summary: "Wade into the crocodile's world with our new book for mid-level readers. An engaging narrative of a crocodile family, captivating images, infographics, milestones, and resources make it perfect for young nature enthusiasts"-- Provided by publisher.
Identifiers: LCCN 2023059424 (print) | LCCN 2023059425 (ebook) | ISBN 9798889891918 (library binding) | ISBN 9781682775646 (paperback) | ISBN 9798889891994 (ebook)
Subjects: LCSH: Crocodiles--Juvenile literature. | Crocodiles--Life cycles--Juvenile literature. | Crocodiles--Australia--Mary River (N.T.)--Juvenile literature.
Classification: LCC QL666.C925 S3354 2025 (print) | LCC QL666.C925 (ebook) | DDC 597.98/2--dc23/eng/20240123
LC record available at https://lccn.loc.gov/2023059424
LC ebook record available at https://lccn.loc.gov/2023059425

CONTENTS

MEET THE FAMILY

SALTWATER CROCODILES

of the Mary River

The Mary River in northern Australia is 140 miles (225 kilometers) long. This region has a dry season and a wet season. Temperatures are warm year-round. Wildlife is plentiful. Azure kingfishers fly overhead. Wallabies visit the riverbank. Large barramundi swim through the dark waters.

This river is also home to the world's largest reptile: the saltwater crocodile. Salties spend time far out in the sea, but they mate and breed in rivers during the wet season.

One February day, a female crocodile came onshore. She built a dirt mound near the water that measured almost two feet high (0.6 meters) and six feet across (1.8 m). There she laid 50 large, leathery, white eggs. She covered them with mud and sticks. For months, she guarded and warmed the nest. She scared off monitor lizards and feral pigs that came near.

Now, on this warm May day, her eggs are about to hatch.

CLOSE-UP

Egg tooth

The egg tooth is not really a tooth. It's a hard bit on the tip of a crocodile hatchling's snout. It helps the hatchling cut through the membrane just inside the eggshell. The egg tooth disappears within a week.

CHAPTER ONE

LIFE BEGINS

Crocodiles are large reptiles. The largest species is the saltwater crocodile of Australia. It can grow to 23 feet (7 m) long and weigh 2,200 pounds (998 kilograms). Every giant crocodile starts out as a small egg. The temperature of a nest determines the gender of the hatchlings that are born there. A nest kept at 91 degrees Fahrenheit (32.8 degrees Celsius) will produce mostly males. A nest that's cooler or warmer than that will produce mostly females. Crocodiles are the only reptiles to care for their young. Mothers stay near their hatchlings for several months to protect them from predators.

All crocodiles have a long snout, strong jaws, many teeth, and a powerful tail. They have tough outer skin with scales and **scutes**. They also have short legs and a long, heavy body. Crocodiles are carnivores. They're not picky eaters, and they will eat whatever meat they can catch. That's often birds and fish and turtles. Big crocs can eat large mammals. Nile crocodiles in Africa sometimes eat zebras and wildebeests. Mugger crocodiles in India eat monkeys.

SALTWATER CROCODILE MILESTONES

DAY 1

- Uses egg tooth to hatch from egg
- Length: 11 inches (28 centimeters)
- Weight: 2.5 ounces (70 grams)
- Pale yellow and black markings
- Fully formed
- Catches own food

Crocodiles are **semiaquatic**. They live on land and in water. On land, they use both a low walk and a high walk. In a low walk, their bellies glide along the dirt. In a high walk, they hold their bellies off the ground and stand tall on all four feet.

Most crocodiles live in tropical regions. Some species prefer fresh water, but many prefer salt water. Most crocs can live in **brackish** waters, which have a mix of the two. A croc's coloring is generally green, gray, or brown. This helps it blend into water and muddy riverbanks.

CLOSE-UP

Teeth

Crocodiles have 50 to 100 teeth. They're good for gripping and tearing, but not for chewing. That's why crocodiles tear their prey apart. Crocs are always growing new teeth. A crocodile might replace each tooth up to 50 times during its life.

FEATURED FAMILY

Welcome to the World

"Urmp. Urmp!" A chirp-honk sound emerges from the nest. The mother crocodile recognizes it. She gently digs away mud to reveal the eggs. One by one, hatchlings use a special egg tooth to poke out. Only about one-quarter of the eggs will hatch. When they emerge, the hatchlings are about 11 inches (28 cm) long. They are pale yellow and black. The mother crocodile gathers her babies in her mouth. She carries them carefully into the water. One curious male hatchling is first to venture boldly into the river.

Crocodiles are masters at not moving at all. They spend a lot of time waiting for prey, mostly while they are submerged under the water. Nile crocodiles in Africa will keep still for hours or even days. Crocodiles are also masters at quick movement. When the right prey comes along, they lunge with amazing speed. The attack is over in a fraction of a second. Then they are still again.

As apex predators, crocodiles are at the top of their food chain. Their hunting keeps prey populations in balance. Crocodile eggs and hatchlings are part of the food web too. They're at the bottom. They provide food for other species and even for other crocodiles. All crocs are also scavengers. That means they'll eat **carrion**, such as dead animals that floods deliver to the river.

(6) WEEKS

- Already aggressive toward other crocs
- Egg tooth is gone
- Stays close to mother
- Mostly eating insects and spiders
- Females and males are the same size

CLOSE-UP

Scutes

Scutes are a special kind of scale with a bony base. They form ridges down a crocodile's back and tail. Scutes help a crocodile move powerfully through water by sweeping its tail back and forth sideways.

FEATURED FAMILY

First Meal

Although the mother crocodile stays nearby for protection, the hatchlings catch their own food. The little male is especially eager. He rests in the shallow water where grasses bend in the breeze. A grasshopper jumps. The small saltie grabs it with tiny, sharp teeth. He gulps the bug down whole. For now, he is happy to eat insects and other invertebrates. Soon, he'll move on to frogs and small fish. Later, he'll progress to much larger prey.

Crocodiles come in ALL SIZES!

Human Male: 6 feet (1.8 m) tall

Congo Dwarf Crocodile: 6.6 feet (2 m) long

Mugger Crocodile: 11.5 feet (3.5 m) long

Nile Crocodile: 13.1 feet (4 m) long

Saltwater Crocodile: 17 feet (5.2 m) long

③ **MONTHS**

- Length: 13.4 inches (34 cm)
- Weight: 3.5 ounces (100 g)
- Mother no longer protecting hatchlings

CLOSE-UP
Nictitating eyelid

This structure is sometimes called a third eyelid. Actually, it's a clear membrane. It slides back and forth across the eye instead of moving up and down. This membrane protects the crocodile's eyes underwater. However, it also makes it harder for a crocodile to focus its vision.

CHAPTER TWO

EARLY ADVENTURES

After a few months, hatchlings are on their own. Their world is dangerous. In some species, only one out of every 100 hatchlings survives to adulthood. Fish, wading birds, turtles, and lizards are threats to young crocodiles. So are larger crocodiles, who see young males as future threats. Once crocodiles reach around three feet (0.9 m) long, they are safe from many dangers.

Crocodiles have several hunting methods. They often **ambush** their prey. With this method, crocs hide and wait for the right moment to launch a surprise attack. Crocodiles also hunt by swimming after prey on the water's surface. Sometimes, they take prey while underwater. A crocodile's third eyelid prevents it from seeing clearly underwater, though. Crocs usually hunt at night when their vision is at its sharpest.

Since crocodiles can't chew their food, they swallow it in large chunks. Luckily, the liquid in their stomachs is highly acidic. It quickly breaks down their food.

6 MONTHS

- Length: 16 inches (40 cm)
- Weight: 8 ounces (227 g)
- Still must avoid predators
- Eats small fish and frogs

CLOSE-UP

Sensory organs

Crocodiles have thousands of bumps on their faces and bodies. These special sensory organs can feel a ripple of movement in the water or a slight change in temperature.

FEATURED FAMILY

Look Who's Sleeping with One Eye Open

The young saltie is six months old and more than a foot long (40 cm). Many animals consider him a snack. Turtles, large fish, pythons, and larger crocodiles are all threats. That's why the young croc sleeps with one eye open! This behavior is called **unihemispheric sleep**. When the left half of the brain shuts down to rest, the right eye closes. One eye and half of the brain remain awake and alert. Then, the sides switch. The strategy helps the little saltie survive. When a hungry barramundi swims too close, he instantly wakes and glides away.

Hunting does not take up all of a crocodile's time. Because they're **cold-blooded**, crocs go ashore to bask in the sun and warm their bodies. If they get too warm, they open their jaws to release heat. Sometimes, in cold or dry periods, crocodiles dig burrows in the riverbank. They crawl in and rest until conditions are better.

Crocodiles live in warm climates. They live mostly in parts of Africa, Asia, and Australia. A few species are found in the Americas. Crocs have salt glands on their tongues that allow them to tolerate salt water. They have been found far from land out in the open ocean.

Wherever they live, crocodiles need both land and water. On land, they nest, eat land animals, and bask in the sun. In the water, they hide, hunt, and mate.

(2) YEARS

- Length: 4.8 feet (145 cm)
- Weight: 26 pounds (12 kg)
- Males now growing larger than females
- Eats small mammals on riverbank

CLOSE-UP

Ears and nostrils

When crocs go underwater, muscular flaps close over their ears and noses to keep water out. This doesn't affect their ability to hear. However, crocodiles don't seem to be able to smell underwater.

FEATURED FAMILY

Give It a Try

At a year old, the saltwater crocodile is about two feet long (61 cm) and weighs four pounds (2 kg). He is learning to be a stealthy hunter. He spots a jacana. The red-feathered bird stands on a lily pad, plucking insects from the water. The saltie submerges. He swims swiftly, but he makes no ripples. This may be because the bony ridges on his back create underwater waves that cancel out surface ripples. Arriving below the lily pad, he waits. The jacana reaches down, and the saltie lunges. He snaps his jaws closed in one-twentieth of a second. Success!

CROCODILES go *ashore* to BASK in the SUN.

(6) YEARS

- Length: 6 feet (1.8 m)
- Weight: 62 pounds (28 kg)
- Greatest danger is from bigger crocodiles

(12) YEARS

- Females reach sexual maturity and mate for the first time
- Females dig practice nests to find the perfect spot for their eggs
- Eats fish, waterfowl, and larger mammals

CLOSE-UP

Fighting

When crocodiles fight, it's a mighty display of strength. Two crocodiles will lunge, bite, and slam their jaws against each other. The loser might escape with its life. Or it might end up down the throat of the winner.

CHAPTER THREE

LIFE LESSONS

Crocodiles are formidable hunters. A croc has many **adaptations** that help it move skillfully through the water. Its eyes, ears, and nose are all located on top of its skull. While it waits to ambush prey, most of the animal's body hides underwater.

Crocs can use their muscles to shift their internal organs around. They can adjust the position of their lungs to make themselves float more easily. This allows them to rise and sink in the water without moving any visible part of their body.

17 YEARS

- Length: 10 feet (3.3 m)
- Can take down water buffalo
- Males reach sexual maturity
- Fights dominant male
- Gets chased out of territory
- Swims through ocean to find new river
- Mates for the first time

Crocodiles do not live alone, but they are territorial. Sometimes, many crocodiles can live peacefully in the same area. However, many species are aggressive. Crocs fight over food and other resources. Males often fight each other for territory. This is especially common during breeding season. Most crocodiles start breeding at 10 to 15 years old. Depending on the species, crocodiles live 30 to 80 years in the wild.

Each section of a water system has one dominant male. It's usually the oldest, biggest crocodile in the area. As other males grow or enter a territory, there will be mighty fights for dominance. Often, the loser of a battle is forced to leave the territory. Sometimes, crocodiles fight each other to the death.

FEATURED FAMILY

This Is How It's Done

At 17 years old, the saltie has become a fierce predator. On a sunny August afternoon, a small herd of water buffalo cools off in the river. The crocodile submerges completely and glides toward them. In an unbelievably quick attack, he clamps his jaws around a buffalo. The saltie rolls over and over in the water, whipping his head and tail. The buffalo drowns. The croc's thrashings tear the buffalo's body apart. The crocodile has perfected this **death roll** as a way to both kill prey and break it down into bite-sized chunks.

CLOSE-UP

Mating

During breeding season, males make bellowing sounds and slap their heads on the surface of the water. Females point their heads upward. When a pair agrees to mate, there is chirping and bubble blowing. Mating takes place completely underwater.

40 YEARS

- Length: 17 feet (5.2 m)
- Weight: 1,000 pounds (454 kg)
- Dominant male of his territory
- Fully grown, but may continue to grow very slowly throughout life

Crocodiles are huge, powerful animals, and people are curious about them. In some areas, crocs attract tourists. In Australia, tourists take river cruises. They want to see saltwater crocodiles up close. Tour operators feed the crocodiles meat, so crocs learn to connect the boats with a good meal. They are sure to gather round and give the tourists a show.

CLOSE-UP

Tails

Crocodile tails have many uses. While swimming, crocs use them like oars. Their tails arch up for threat displays. Some crocs use their tails to help them push off powerfully from the bottom of the river.

FEATURED FAMILY

Practice Makes Perfect

The male crocodile is now 11 feet (3.4 m) long. He is ready to mate. It will be a major challenge. Salties are territorial. A dominant male rules each section of the river. When the saltie dares to approach a female croc, the dominant male in the area glides over. He rises up and uses his powerful head to smash the smaller crocodile. The younger saltie knows he is no match. He wriggles free and swims away to try a different area. He might even swim out to sea and then into another river system. With persistence, the saltie may someday be a dominant croc with his own territory.

Male crocodiles have MIGHTY *fights for* DOMINANCE.

73 YEARS

- Scarred from years of fighting
- Has difficulty defending territory
- End of life

CHAPTER FOUR

HELPING CROCODILES SURVIVE

Archosaurs were a group of reptiles that lived 250 million years ago. Dinosaurs descended from them. Both archosaurs and dinosaurs went **extinct**. Today, only two descendants of archosaurs survive: birds and crocodiles. So, crocodiles' closest living relatives are birds, and their closest extinct relatives are dinosaurs.

Will crocodiles go extinct too? Of 17 living crocodile species, five are critically endangered. Two more are vulnerable.

What is killing off crocodiles? Habitat loss is the main threat. As humans build and develop land, there is less room for wild animals. People change water systems. Building dams and rerouting waterways can stop crocodiles—or their prey—from reaching parts of their habitat. Illegal hunting, fishing nets, and water pollution are also problems for crocodile populations.

Crocodiles are not cute or cuddly. Some species even attack and eat people. It can be hard to make the public care about crocodile survival. But crocodiles are worth protecting. In the 1970s, saltwater crocodiles were almost extinct. New laws were passed to make it illegal to hunt and trap crocs. Today, salties have a stable and healthy population.

The species most in danger is the Philippine crocodile. Scientists believe fewer than 150 remain in the wild. The Mabuwaya Foundation in the Philippines is one group working to save them. They promote fishing methods that don't harm crocodiles. They collect hatchlings and raise them until they're old enough to have a good chance of survival. Then they release them into the wild. They also try to educate the public about the importance of crocodiles.

Crocodiles have been on Earth for a very long time. They're an important part of many ecosystems and food webs. They also have cultural and religious importance to some groups of people. Our world wouldn't be quite as wild and wonderful without the impressive, sometimes aggressive, crocodile.

FAMILY ALBUM

SNAPSHOTS

American crocodiles dig long burrows into riverbanks. A burrow can be almost 30 feet (9.1 m) long. It has a big area at the end to turn around in.

Freshwater crocodiles in northern Australia are the fastest crocodiles on land. They've been recorded galloping 11 miles (18 km) per hour.

Philippine crocodiles nest in the mountains of Luzon, Philippines. There, fast-flowing rivers can sweep away and drown hatchlings.

Morelet's crocodiles have very broad snouts, making them look more like alligators than crocodiles.

Saltwater crocodiles are the largest reptiles in the world. They are sometimes considered the animal most likely to eat a human.

In the 1900s, hunting reduced the number of **Orinoco crocodiles** from three million to only a few hundred individuals.

In some communities, people celebrate **West African crocodiles** as bringers of rain.

Experts believe that 200 to 300 people are eaten each year by **Nile crocodiles**. They're often people who depend on rivers for cooking and bathing.

New Guinea crocodiles are farmed for their skin. The government allows this practice so that people don't hunt wild crocodiles instead.

Mugger crocodiles of India get their name from the Hindi word *magar*, which means "water monster."

Cuban crocodiles have striking yellow, black, and white markings.

In Central Africa, **dwarf crocodiles** face huge threats from hunters who need their meat to eat or who sell the meat to make money.

WORDS to Know

adaptations	changes that a living thing has gone through over time to make it better suited to its environment
ambush	to attack by surprise from a hidden place
brackish	containing both salt water and fresh water
carrion	dead, rotting meat
cold-blooded	having a body temperature that changes according to the temperature of the surroundings; reptiles and fish are cold-blooded
death roll	the technique of rolling and thrashing that most crocodiles use to drown and break up their prey
extinct	no longer existing; died out
invertebrates	animals without spines, such as insects, spiders, and snails
scutes	bony or large scales
semiaquatic	living mostly, but not completely, in water
unihemispheric sleep	sleep that involves only half of the brain and one eye at a time, allowing the other half to stay alert

LEARN MORE

Books

Davey, Owen. *Curious About Crocodiles*. London: Flying Eye Books, 2021.

Shields, Jennifer. *Alligator or Crocodile? A Compare and Contrast Book*. Mount Pleasant, S.C.: Arbordale Publishing, 2023.

Taylor, Charlotte. *Crocodiles: Ambush Killers*. Berkeley Heights, N.J.: Enslow Publishing, 2021.

Websites

"American Crocodile: Species Profile." National Park Service. https://www.nps.gov/ever/learn/nature/crocodile.htm

"Crocodilian." San Diego Zoo. https://animals.sandiegozoo.org/animals/crocodilian

"What on Earth is this crocodile doing?" Australian Geographic. https://www.australiangeographic.com.au/topics/wildlife/2020/09/what-on-earth-is-this-crocodile-doing-a-crocodile-expert-explains

Documentaries

Beman, Jennifer. *Crocodile King*. National Geographic, 2010.

Davey, Joel. *Secrets of Wild Australia: The Crocodile*. London: Blink Films, 2016.

Mackinder, Simone. *Crocodiles, Alligators, Caimans & Gharials*. Screen Queensland, 2018.

Note: Every effort has been made to ensure that any websites listed above were active at the time of publication. However, because of the nature of the Internet, it is impossible to guarantee that these sites will remain active indefinitely or that their contents will not be altered.

Visit

FLAMINGO MARINA

Spot salt water-loving American crocodiles in the wild at this marina in Everglades National Park.

1 Flamingo Lodge Hwy.
Homestead, FL 33034

INDIANAPOLIS ZOO

Check out a rare Orinoco crocodile in an exhibit called Alligators & Crocodiles: The Fight to Survive.

1200 W. Washington St.
Indianapolis, IN 46222

ST. AUGUSTINE ALLIGATOR FARM

This zoological park is home to Nile crocodiles and many other crocodilian species. Exhibits, daily shows, and daily feedings.

999 Anastasia Blvd.
St. Augustine, FL 32080

TOLEDO ZOO AND AQUARIUM

Visit Baru, the resident saltwater croc.

2 Hippo Way
Toledo, OH 43609

INDEX